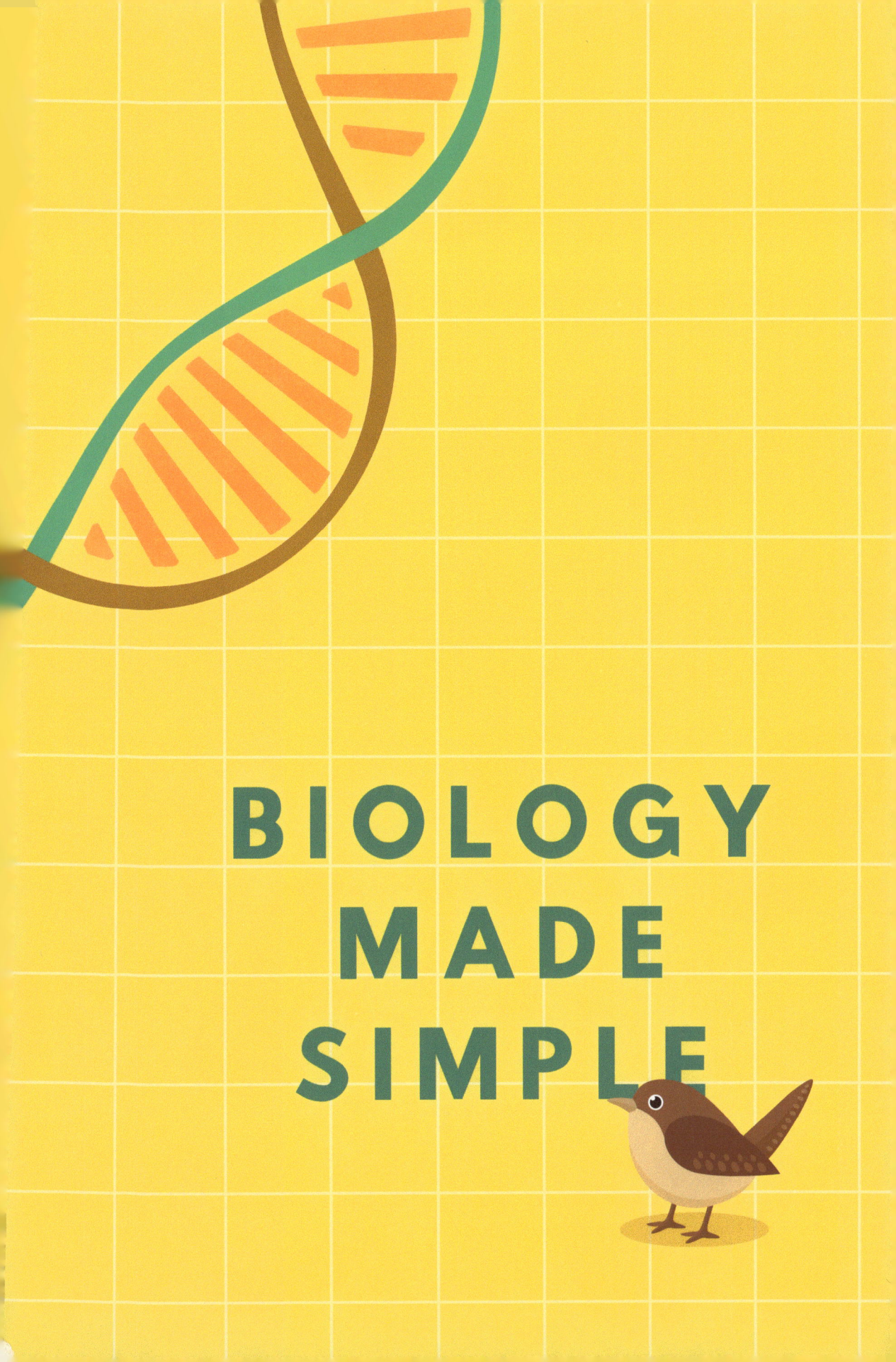

BIOLOGY MADE SIMPLE

A gateway introduction EVERY young student should read

**Written and Designed by
RION IWANO**

First published May 2025
Copyright © 2025 Rion Iwano

All rights reserved.
No part of this publication may be reproduced, stored in a retrieval system, or transmitted in any form, or by any means (electronic, mechanical, photocopying, recording or otherwise) without prior written permission from the publisher.

ISBN 978-1-7641014-0-0 (Paperback)

ISBN 978-1-7641014-1-7 (Hardcover)

ISBN 978-1-7641014-7-9 (eBook)

A catalogue record for this book is available from the National Library of Australia

Disclaimer:
The material in this publication is intended as general information only and does not constitute professional advice. It is not intended to provide specific guidance for particular circumstances. Readers should obtain professional advice where appropriate. To the maximum extent permitted by law, the author and publisher disclaim all responsibility and liability for any loss, damage, or consequences arising from the use of, or reliance on, the information contained in this publication.

To my favourite teachers who inspired me to love science and biology :)

TABLE OF CONTENTS

What is Biology?............8
A Living Thing............... 9

What are Cells?............. 10
Taxonomy...................... 13
Kingdoms...................... 14
Types of Animals........... 16
Ecosystems................... 18
Genetics....................... 20
How DNA is Read.......... 22
How Cells Divide........... 24

The Human Body............ 25
The Brain...................... 28
Nervous System............. 30
Digestive System.......... 32
Respiratory System....... 34
Cardiovascular System.. 36

Sight............................. 38
Hearing......................... 39
Smell............................. 40
Taste............................ 41
Touch............................ 42

Homeostasis................ 43
Disease......................... 44
Immune System........... 46
Vaccines....................... 48
Antibiotics.................... 49

What are Plants?......... 50
Photosynthesis............ 51
Plant Structure............ 52
Leaf Structure............. 53
Pollination.................... 54

Inheritance.................. 56
Adaptations................. 58
Evolution...................... 60

WHAT IS BIOLOGY?

Biology is the **study of living things** – from tiny bacteria to the largest animals and plants. At its core, biology explores how living things are built, how they work, and how they interact with each other and their environment.

There are many different branches of biology. Some focus on specific types of organisms, like **microbiology** (the study of microbes), **botany** (the study of plants), and **zoology** (the study of animals). Other scientists study the structure of the human body, which is called **anatomy**, or how living things interact with their surroundings, known as **ecology**. Biology also connects with other sciences. For example, **bioengineering** brings together biology and engineering to create new solutions and technologies.

LIVING THINGS

So, what is a **living thing**? What makes a human a living thing, but not a robot? What makes a plant alive, but not a book?

Living things, also called **organisms**, share certain features, or **criteria**, that set them apart from non-living things:

Movement

Respiration

Sensitivity

Growth

Reproduction

Excretion

Nutrition

For example, a sunflower is a living organism because it **moves** toward the light, **respires** to produce energy, is **sensitive** (responsive) to environmental changes (such as light), **grows** from a seed into a plant, **excretes** waste products like carbon dioxide, and requires **nutrition** from the soil and sunlight.

 **"MRS GREN"** is a helpful acronym to remember the criteria for living things

WHAT IS A CELL?

Cells are the **basic building blocks** of all **living things**, that is to say, all living things are made of cells. Cells perform important functions, such as converting nutrients into energy.

Each cell is surrounded by a **membrane** and is made up of different parts, depending on the type of organism. The diagram below shows the main parts of an **animal cell**:

CELL MEMBRANE
Separates the cell from its environment

NUCLEUS
Contains genetic material called **DNA**

ORGANELLES
All the structures within the cell are called organelles, and each performs a specific function. For example, the **mitochondria** converts nutrients into usable energy.

Plant cells are similar to animal cells – they both have a cell membrane, nucleus and organelles like mitochondria. However, plants also have some additional structures which help the plant to survive. For example, plants have a **cell wall** for support, as well as unique organelles such as **chloroplasts**, which are the site for **photosynthesis** (the process by which plants convert sunlight into energy – see page 51 for more).

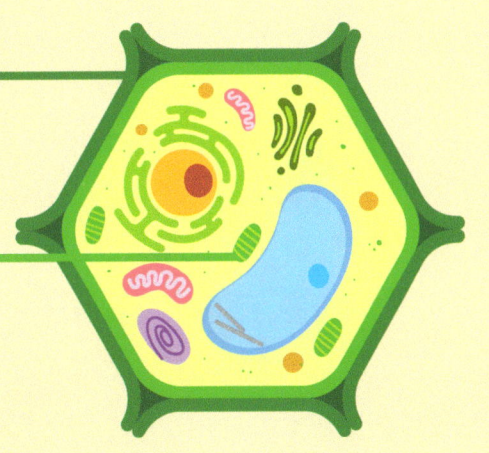

Cells that have organelles that are **bound by membranes** are called **eukaryotic**. Plants and animals (including humans) are both eukaryotic.

However, not all living things are eukaryotes. Cells that lack distinct membrane-bound organelles are **prokaryotes**. An example of a prokaryote is bacteria.

Unlike eukaryotes, prokaryotes don't have a nucleus. DNA that floats in a **nucleoid** region

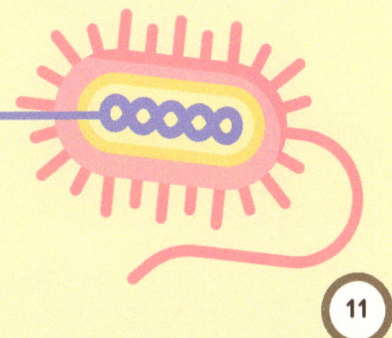

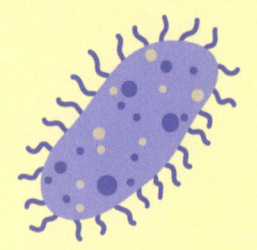 Some organisms, such as bacteria, are only made of one cell; these are called **unicellular**.

Other organisms like animals and plants are made up of **trillions** of cells and are **multicellular**. In multicellular organisms, cells are **specialised**. For example your red blood cells and nerve cells have **differentiated** to perform specific functions.

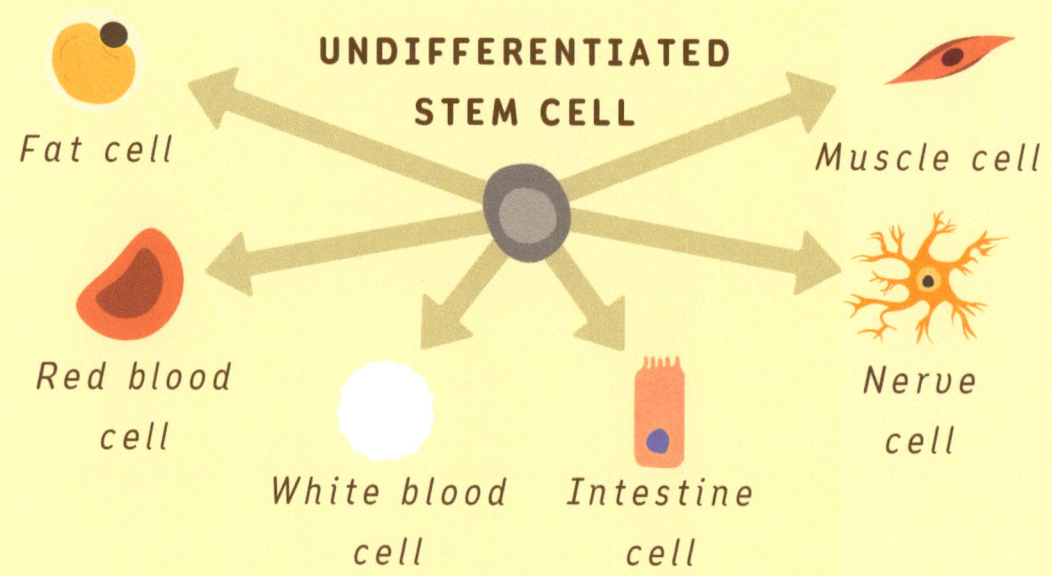

These specialised **cells** form **tissues**, which form **organs**, which form **organ systems**, which together make up **organisms**.

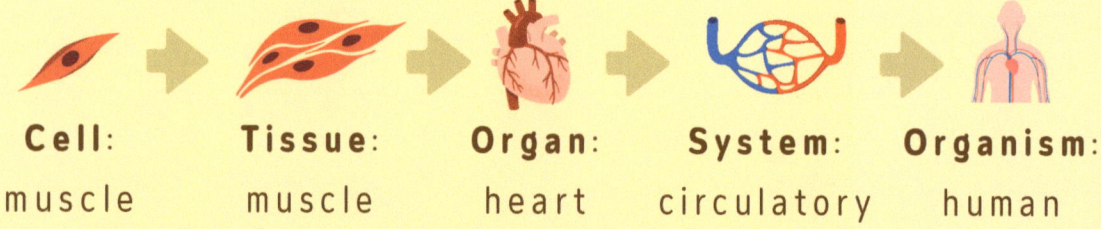

TAXONOMY

Scientists use a system to describe and classify living things, known as **taxonomy**. Organisms are grouped into a hierarchical system of classification. This system includes three **domains**, which are divided further into **kingdom, phylum, class, order, family, genus, and species**.

For example, the emperor penguin belongs to the domain **Eukaryota**, kindom **Animalia**, and so on. Its species name is derived from its **genus** and **species** names, which in this case is *Aptendotyes forsteri*.

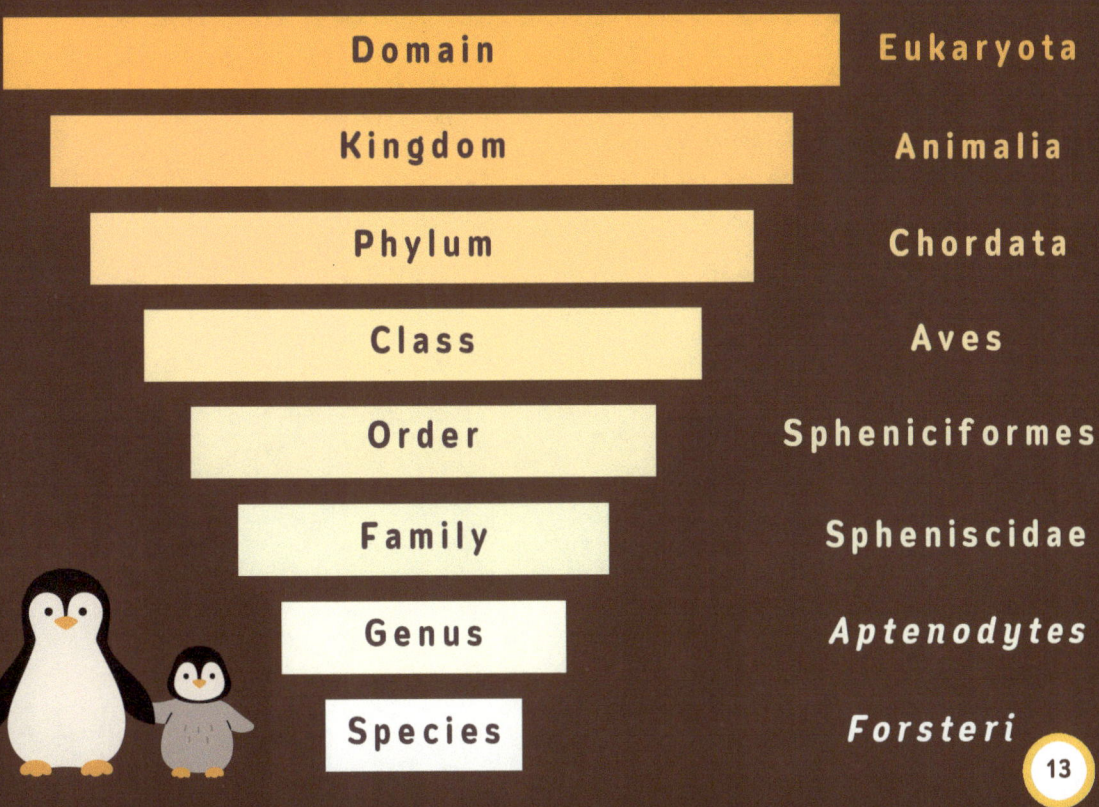

Domain	Eukaryota
Kingdom	Animalia
Phylum	Chordata
Class	Aves
Order	Sphenisciformes
Family	Spheniscidae
Genus	*Aptenodytes*
Species	*Forsteri*

KINGDOMS

In the 1700s, there were two main kingdoms - **animals** and **plants**. However, with more advanced technology, scientists now classify living things into **six distinct kingdoms**.

ANIMALS

Animal organisms are multicellular eukaryotes, like budgies. They depend on plants and other organisms for nutrition.

PLANTS

Plants are also multicellular eukaryotes. However, they photosynthesise to convert energy from the sun into usable energy.

FUNGI

Fungi are eukaryotic and may be multi- or unicellular. They obtain nutrition by decomposing organic matter. Examples include mushrooms, moulds and yeast.

PROTISTA

Protists are unicellular eukaryotes. They encompass all other organisms that are not classified Animalia, Plantae or Fungi, such as amoebas.

EUBACTERIA

Eubacteria are "true bacteria". They are unicellular prokaryotes. They can be harmful, like *Salmonella*, or beneficial, such as those that aid digestion.

ARCHAEBACTERIA

These are also unicellular prokaryotes. Archaebacteria are unique because they can survive in extreme environments, such as hot springs.

Each of these kingdoms can be further classified into groups. For example, **animals** can be classified based on the types of food that they consume (**herbivores**, **carnivores**, **omnivores**). See the next page for more information on animal classification.

ANIMAL TYPES

Animals can be distinguished based on various types of characteristics.

One characteristics used is whether the animal is **warm-blooded** and can maintain their own body heat (**endotherms**) or **cold-blooded** and rely on their environment to maintain body temperature (**ectotherms**).

Another example is whether they have a **backbone** (**vertebrates**) or do not (**invertebrates**).

There are six major groups used to classify animals.

BIRDS

Birds are endothermic vertebrates with feathers, wings and beaks. They lay hard-shelled eggs.

MAMMALS

Mammals are endothermic vertebrates with hair or fur. They nourish their young with milk.

FISH
Fish are vertebrates that live in water and breathe using gills.

REPTILES
Reptiles are ectothermic vertebrates with dry skin covered in scales or bony plates. They typically lay soft-shelled eggs.

AMPHIBIANS
Amphibians are ectothermic vertebrates without scales. They spend part of their lives in water and part on land.

INVERTEBRATES
Invertebrates are animals without a backbone. They include insects, jellyfish, and crustaceans.

ECOSYSTEMS

An ecosystem is made up of all the **biotic** (living) and **abiotic** (non-living) things in a particular area.

Every organism in an ecosystem has a specific role in the **food chain** – as a **producer**, **consumer**, or **decomposer**.

Green plants are **producers** because they make their own food through **photosynthesis**.

Animals, including humans, are **consumers** because they **eat** plants and/or other animals.

Decomposers, such as **bacteria** and **fungi**, **break down** dead plants, animals, and waste products. This process returns important nutrients back into the soil, helping new plants grow.

DID YOU KNOW?
Although rainforests make up less than 2% of Earth's surface, they house more than 50% of the world's plants and animal species.

FOOD WEBS

Food webs consist of multiple **food chains**, showing the interconnected **predator-prey relationships** of an ecosystem. It shows how changes in one species can affect many others. For example, an increase in the grasshopper population may lead to fewer plants and more birds.

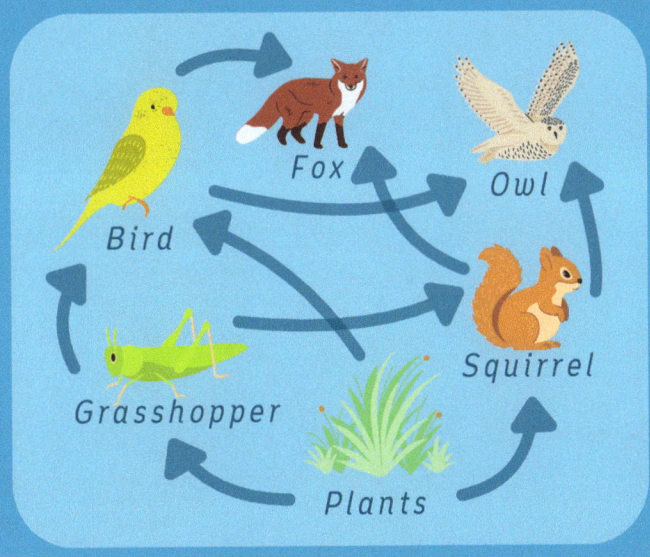

TROPHIC LEVELS

Trophic levels are steps in a **food chain** that show how **energy and nutrients flow** through an ecosystem.

Trophic level

- 5th (Quaternary consumers)
- 4th (Tertiary consumers)
- 3rd (Secondary consumers)
- 2nd (Primary consumers)
- 1st (Producers)

Energy flows through an ecosystem, from the bottom trophic level to the top. Only about 10% is passed on at each step – the rest is lost mainly as heat during an organism's life.

GENETICS

Our genetic information is stored in **DNA**, which stands for **deoxyribonucleic acid**. DNA is a long, thin molecule which acts like a recipe holding the **instructions** that tell our bodies how to develop and function.

Cell

For eukaryotes, DNA is located in the **nucleus** of each cell. The DNA is packaged into structures called **chromosomes**.

Chromosomes

DNA

All humans have **23 pairs of chromosomes** for a total of **46**. Other organisms like cats have two sets of 19 for a total of 38. Most organisms have two sets of chromosomes – one from the **mother** and one from the **father**. These organisms are called **diploid**. Organisms that only have one set are called **haploid**.

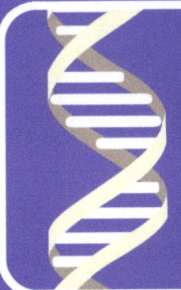 When you look at DNA under a really strong microscope, it looks like a **twisted** ladder made of **two strands**. Scientists call this shape a **double helix**.

The strands are made up of a sequence of bonded **nucleotides** (sometimes referred to as "bases").

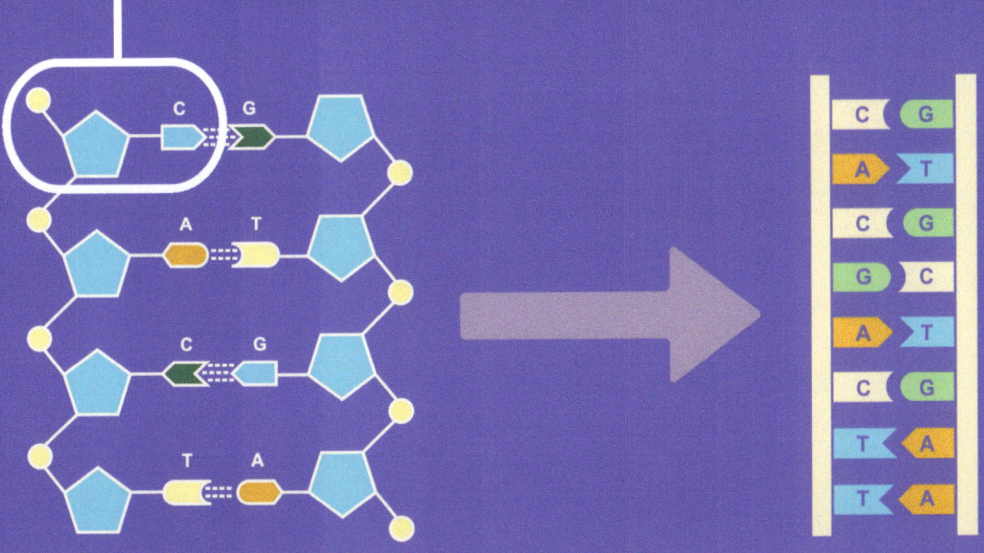

DNA has four types of nucleotides: **adenine (A)**, **thymine (T)**, **cytosine (C)**, and **guanine (G)**. Their order acts like a **code** that tells the body what to do, such as which **proteins** to make (see the next page to learn about how proteins are made). **Genes** are **sections of DNA** or groups of nucleotides. Genes help decide **traits** like height or eye colour.

Sometimes mistakes happen when DNA is copied. These mistakes are called **mutations**.

HOW DNA IS READ

The human body is made of thousands of different **proteins**; in fact around 20% of our body is composed up of proteins. They are essential for a wide range of functions, such as hormone regulation and immune function. Proteins are made of long chains of **amino acids**.

So, how is the DNA sequence read to create proteins?

This is done through a process called **protein synthesis**.

1. TRANSCRIPTION

Here, a copy of the **DNA** sequence is made.

This copy is called **messenger RNA**, as it carries genetic information to the next step.

It is similar to DNA, but it is a single strand and is made from a substance called ribonucleic acid.

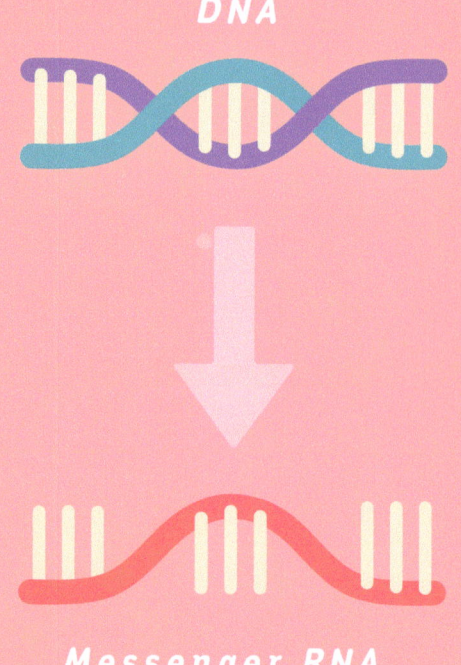

2. TRANSLATION

The **messenger RNA** exits the nucleus and attaches itself to a **ribosome**.

At the **ribosome**, the sequence of nucleotides is read in groups of **three**, called **codons**. Each codon codes for an **amino acid.** For example, the sequence "**AGC**" codes for the amino acid **serine**.

The ribosome moves down the **messenger RNA** until it reaches a **stop codon**, which ends **translation**.

The resulting product is a **chain of amino acids**. This chain **folds** into three-dimensional shapes and your **protein** is complete.

These completed proteins can be used for various functions, such as **structure** (e.g., collagen found in cartilage) and for **enzymes** (a protein that can speed up chemical reactions).

HOW CELLS DIVIDE

How do we get new cells? Cells are made from other cells through division in a process called **mitosis**. A **parent cell's DNA** is replicated and pulled to opposite ends of the cell. Then, the cell splits into two, giving rise to **two identical daughter cells**.

Before mitosis, DNA is replicated during **interphase**.

Mitosis begins with **prophase**, where DNA condenses into **chromosomes**.

Then the chromosomes line up during **metaphase**.

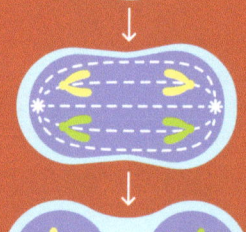
And are pulled apart during **anaphase**.

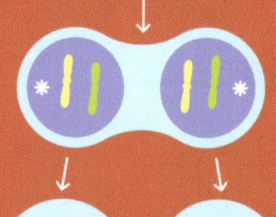
The nuclear membrane reforms during **telophase**.

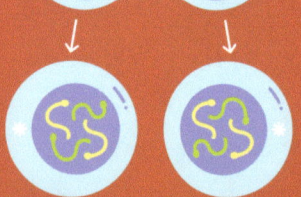
Mitosis ends, and the cell undergoes **cytokinesis**, resulting in 2 new cells.

THE HUMAN BODY

The human body is made up of **206 bones, 78 organs**, and **enough blood vessels to circle the Earth twice**.

The human body is like a supermachine. It has many different parts and **systems**, each with their own job, but also working together to keep together to keep you **alive** and **healthy**.

> **DID YOU KNOW?**
> Whilst most of our body structures serve a function, some are just leftover parts from our evolutionary past. These are called **vestigial structures**.
>
> For example, our tailbone is a remnant of a tail, which humans no longer have.

The body is resilient, but it needs proper care to function well. Eating nutritious **food, exercising**, getting enough **sleep**, and **managing stress** help keep all the systems in balance. Taking care of your body now can also have lasting effects on your **health** later in life.

SKELETAL SYSTEM

Your **bones** protect your organs. They also work with **muscles** to help you move.

RESPIRATORY SYSTEM

Oxygen reaches your **lungs** for gas exchange. You breathe out **carbon dioxide**.

ENDOCRINE SYSTEM

Glands produce **hormones**, which send signals to and regulate organs.

NERVOUS SYSTEM

Your **brain** communicates via **nerves cells**.

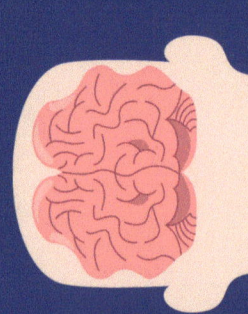

CARDIOVASCULAR SYSTEM

Your **heart** pumps **blood** through **vessels**. Blood carries **oxygen** and other chemicals.

DIGESTIVE SYSTEM

The **stomach** and **intestines** digest food. The body absorbs **nutrients** and removes **waste**.

URINARY SYSTEM Your **kidneys** filters blood and removes waste through a fluid called **urine**.

REPRODUCTIVE SYSTEM This system includes organs that help make babies.

MUSCULAR SYSTEM **Muscles** are responsible for **movement**.

IMMUNE SYSTEM **Lymph vessels** protect against disease.

27

THE BRAIN

Despite only being the **size of two fists** put together, the brain is arguably the most important organ. It is where we do our **thinking** and where we tell the rest of the body what to do.

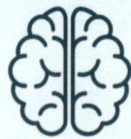

The brain has **two halves**. Because nerves cross over as they enter the brain, the **left half controls the right side** of the body, and the right half controls the left.

The brain has three main parts:

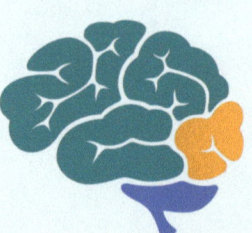

CEREBRUM: controls your voluntary muscles

CEREBELLUM: controls balance and movement.

BRAIN STEM: controls automatic functions like breathing.

The brain is protected by the **skull** and by membrane coverings called the **meninges**. A fluid called **cerebrospinal fluid** surrounds the brain (and spine). It acts as a shock absorber and cushion. It also provides nutrients to the brain and removes waste.

The brain is made up of **86 billion** neurons. These form **100 trillion** connections to each other

Have you ever gotten a **brain freeze** after eating ice cream? The brain itself doesn't have **pain receptors**. Instead, brain freeze occurs when the cold ice cream touches the roof of your mouth. This causes **blood vessels** above your mouth to open and close quickly to regulate your body temperature. The swollen blood vessels push against **pain receptors** above your mouth, and your brain interprets this as pain.

Your brain weighs around **1.4 kg** or 3 pounds.

NERVOUS SYSTEM

The nervous system helps the brain communicate with the rest of the body and informs the brain about what's happening, using **electrical signals**.

The system is made up of the **central** and **peripheral nervous system**. The brain and spine make up the **central nervous system**.

The **brain** is the control centre, and the **spine** is the major highway to and from the brain. The **peripheral nervous system** includes the **neurons** that run throughout the whole body.

Some nerves are **sensory nerves**. These send signals from the eyes, ears, nose, tongue, and skin to the brain. Other nerves are **motor nerves**. These carry messages from the brain to the rest of the body to tell our muscles to move.

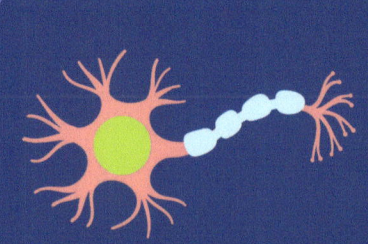

Each neuron is made up of three main parts. **Dendrites** pass signals from other neurons to the **cell body**. The **axon** carries the signal away to other neurons or body tissues, such as muscles.

The things we **actively control**, like talking with our mouths, are controlled by the **somatic nervous system**. Other involuntary actions are controlled by our **autonomic nervous system (ANS)**. These include functions such as:

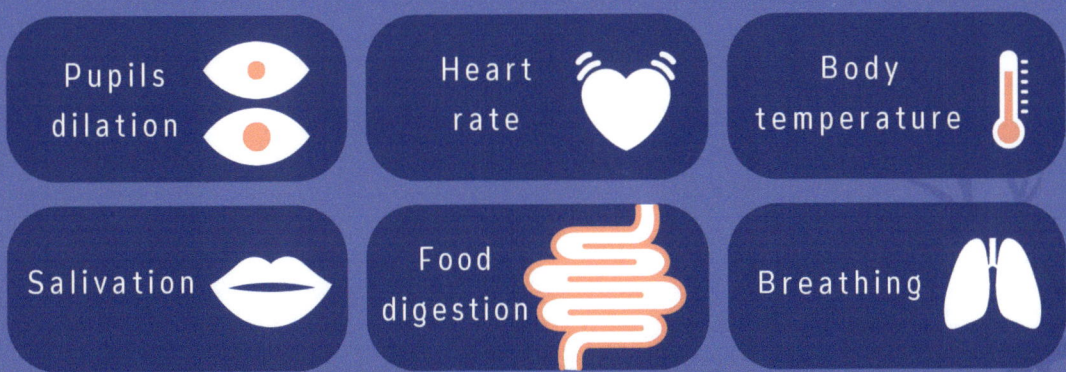

Often, people talk about the five senses. But you actually have more than twenty. For example:

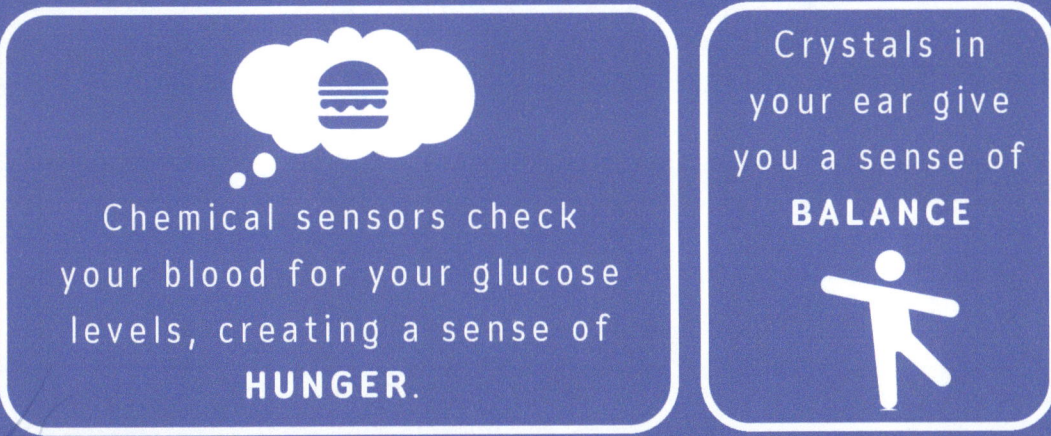

Sometimes our bodies react before our brains. These are called **reflexes**. For example, if we touch something hot, our hand moves away before the brain fully processes it.

DIGESTIVE SYSTEM

The digestive system breaks down food so that we can absorb its **energy**, **vitamins**, and **minerals**. Digestion can be broken down into four main steps.

 **CHEWING**

When you chew your food, it breaks it into **smaller pieces**. Your saliva also contains proteins called **enzymes** that speed up chemical reactions to start breaking down **starchy** foods like rice.

 **STOMACH**

Once you swallow, the food travels down a stretchy tube called the **oesophagus** and enters the **stomach**. The stomach contains **enzymes** that break down substances like protein, as well as **digestive acid** that further breaks down the food and kills harmful **bacteria**.

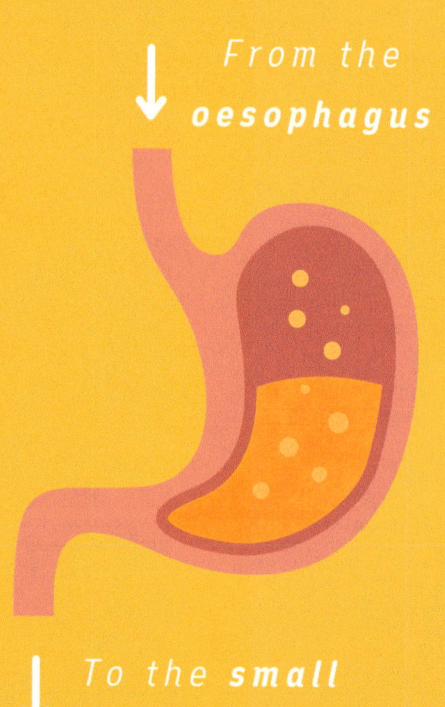

*From the **oesophagus***

*To the **small intestine***

Carbohydrates are digested quickly, whereas **proteins** and **fats** stay in the stomach for longer because they take more time to break down. This is why eating fats and proteins keeps us feeling full for longer than eating carbohydrates alone.

> **DID YOU KNOW?**
> Your stomach would eat itself if it weren't lined with mucus. Mucus is a slimy substance that acts as a protective barrier inside your nose, lungs, stomach, and intestines. Mucus in the stomach helps protect the lining from digestive acids.

 **SMALL INTESTINE**

The **liver** and **pancreas** release **digestive juices** into the **intestine** to further break down carbohydrates, proteins, and fats. These are absorbed from the intestine and into the bloodstream.

 **LARGE INTESTINE**

Any remaining material is broken down by **bacteria**. The waste collects in the **rectum** and leaves the body as **faeces**.

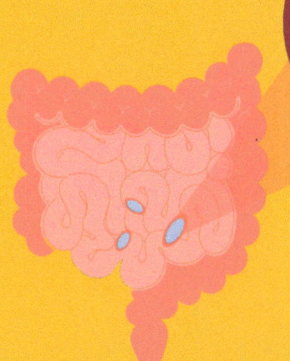

RESPIRATORY SYSTEM

Respiration is the process of breathing in **oxygen** and breathing out **carbon dioxide**. We need to breathe because our cells use oxygen to convert **sugar** into **energy** through the following reaction:

$$\text{sugar} + \text{oxygen} \rightarrow \text{carbon dioxide} + \text{water} + \text{energy}$$

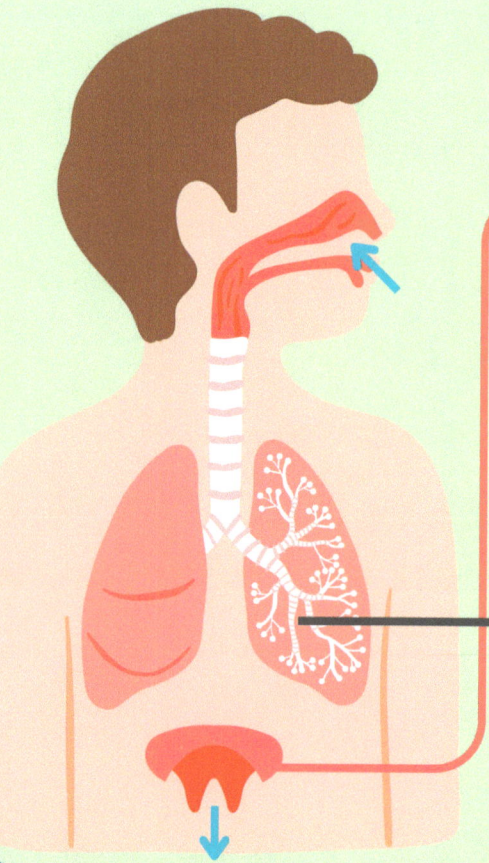

We breathe using a muscle called the **diaphragm**. When it flattens out, our **lungs** expand and fill with air.

The air gets forced through our mouth or nose, down the **windpipe** and into **bronchi tubes**. The bronchi tubes branch out like a tree to distribute air throughout the lung.

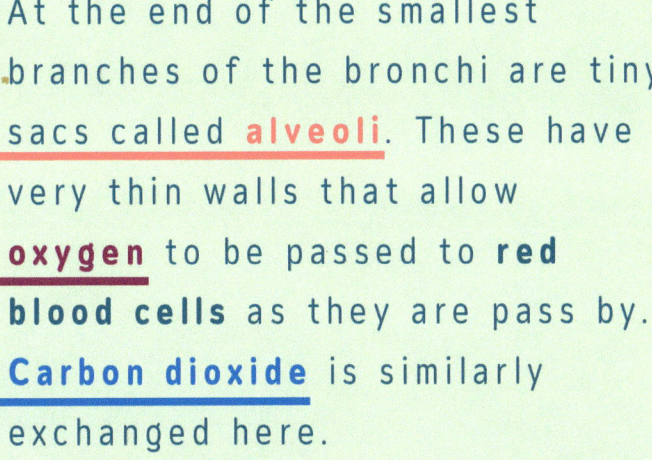

At the end of the smallest branches of the bronchi are tiny sacs called **alveoli**. These have very thin walls that allow **oxygen** to be passed to **red blood cells** as they are pass by. **Carbon dioxide** is similarly exchanged here.

DID YOU KNOW?
Breathing a sigh of relief prevents lung failure. Alveoli collapse over time, which may lead to lung failure. Sighing helps re-inflate collapsed alveoli. This is part of why we sigh 12 times an hour.

Our nose plays an important role in this system. It has tiny hairs, called **cilia**, and **mucus** to filter out dust, germs, and other particles. The nose also warms up the air before it reaches our lungs.

WHY DO WE RUN OUT OF BREATH?
When you do strenuous exercise, your body needs more energy, which means it also needs more oxygen. Your heart pumps faster to circulate more blood through the lungs, and your lungs breathe faster to try to take in more oxygen, making you feel out of breath.

CARDIOVASCULAR SYSTEM

The cardiovascular system includes the heart, blood vessels, and blood. The heart pumps blood through the blood vessels to carry **oxygen** and other **nutrients** to cells throughout the body.

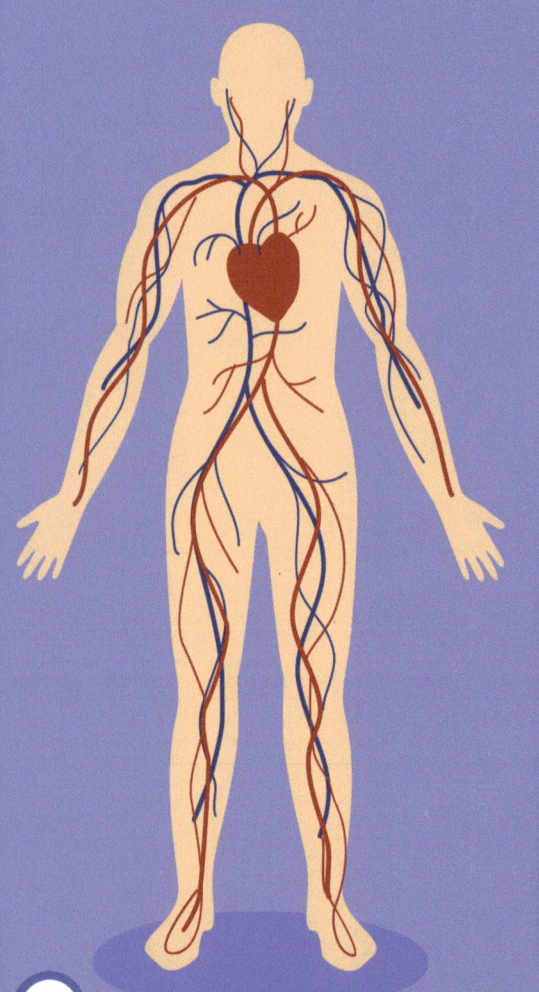

There are two main types of blood vessels: arteries and veins. **Arteries** carry blood from the heart to the rest of the body. This blood contains lots **of oxygen**. **Veins** carry blood back to the heart after the oxygen has been used.

Veins have flap-like structures called **valves** that prevent blood from going the wrong direction.

Every day, your heart pumps enough blood to fill **40 Baths**

Most blood cells are made in **bone marrow** - a spongy tissue that fills bones.

There are two types of blood cells: **red blood cells**, which carry oxygen, and **white blood cells**, which are part of the immune system (see page 46). These cells are carried in a watery fluid called **plasma**.

Red blood cells are shaped like **biconcave discs**. This is because this shape maximises the **surface-area-to-volume ratio**, allowing the cells to carry more oxygen.

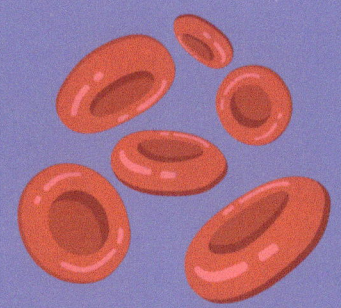

Blood also has functions in other body systems. For example, blood carries away **nutrients** from the digestive system to the rest of the body.

The **spleen** and **liver** clean the blood by removing **old blood cells** and **bacteria**. The **kidneys** filter water, minerals, and other waste from the blood.

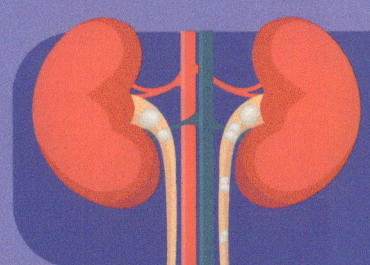

The kidneys process 200 litres of fluid each day. 2 litres become urine, and the rest is returned to the body.

SIGHT

Sight is one of our most important senses. Our **eyes** take in **light** and convert it into **electrical signals** that the **brain** can interpret.

PUPIL
This is where the **light** enters. It **dilates** to let in more light when it is dark.

RETINA
Here, light is processed by **rods** and **cones**, which send signals to the **brain**.

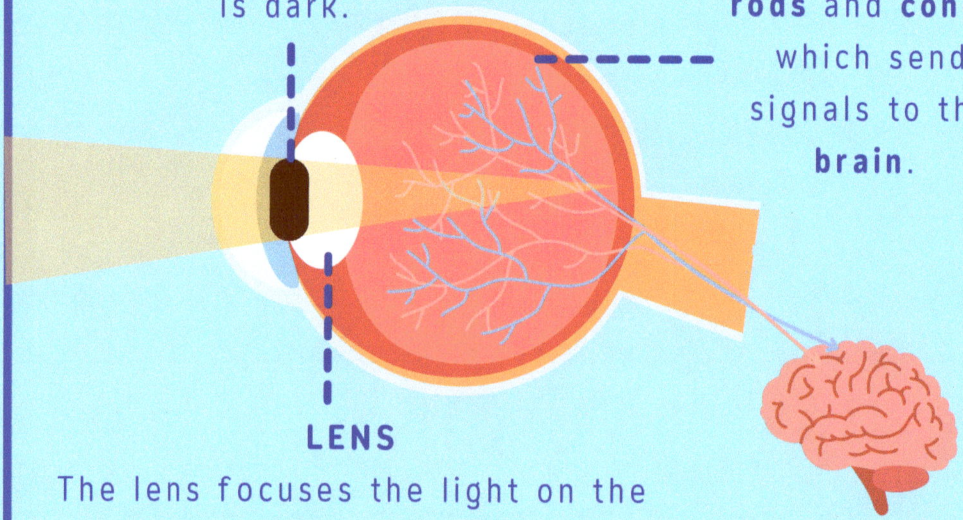

LENS
The lens focuses the light on the **retina** by adjusting its shape.

RODS are extra sensitive to light and help us see in the **dark**.

CONES help us to see **colour**. There are three types of cones: red, green, and blue.

HEARING

Sound travels in the form of waves. The **sound waves** enter your **ear canal** and make your **eardrum** vibrate. The **eardrum** is linked to three bones in your middle ear: the **malleus, incus,** and **stapes**.

These bones amplify and transfer the vibrations to the **cochlea**. The fluid inside the cochlea passes the vibrations onto the thousands of **tiny hair cells**, which translate them into **electrical signals** for the nerve to send to the brain.

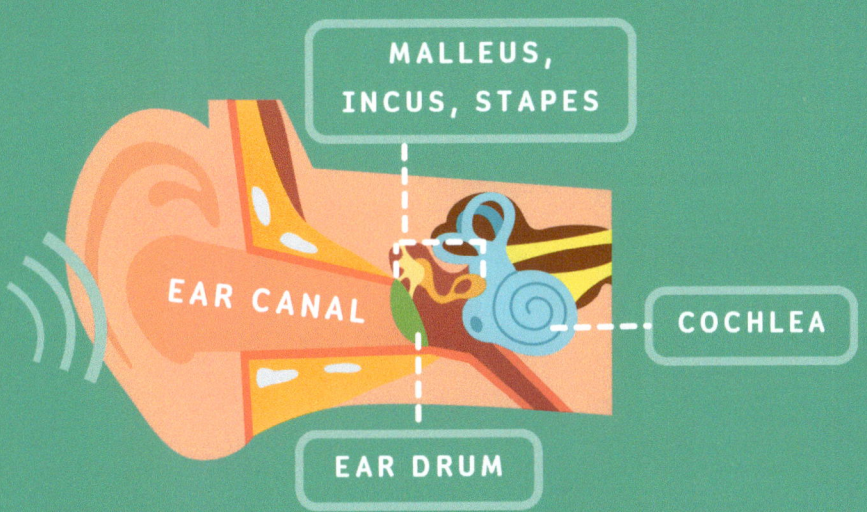

DID YOU KNOW?
Different animals have different hearing **frequency** ranges. Dolphins can hear higher-pitched sounds than we can, but can't hear low-pitched sounds like we can.

SMELL

A batch of cookies in the oven. The salty smell of the ocean breeze. How do you smell these smells, and thousands more?

We smell things that release **odour molecules** that float in the air. When we breathe in, the air, which contains odour molecules, fills our **nasal cavity**. At the back of this region is the **olfactory epithelium**, which contains **receptor cells**. Odour molecules stick to these receptor cells, sending signals to the **olfactory bulb** and then to the **brain**.

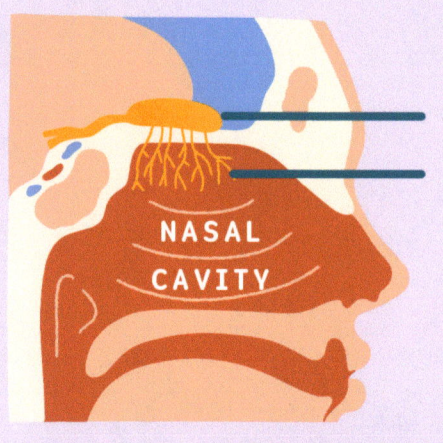

Olfactory bulb
Olfactory Cilia (covered with **receptors**)
NASAL CAVITY

DID YOU KNOW?
Your sense of smell gets better as the day goes on.

HOW DO WE DISTINGUISH BETWEEN SMELLS?

We have 400 unique olfactory receptors. Odour A might trigger receptors 1, 3, and 119, while odour B might trigger receptors 5, 16, and 327. These combinations allow you to detect **1 trillion odours**.

TASTE

Our tongue can taste five main flavours: **sweet**, **sour**, **salty**, **bitter**, and **umami** (a meaty flavour). So, how does your tongue do it?

Your tongue is covered in small bumps that you can see called **papillae**. Papillae have **touch** sensors, **temperature** sensors, and **taste buds**.

Taste buds are tiny organs that have hairs called **microvilli**. These hairs come into contact with the molecules from the food we eat and send messages to your **brain**, telling you what you're tasting.

TONGUE   PAPILLAE   TASTE BUD

Ever wondered why food tastes bland when you're sick? It's because flavour also relies on **smell** – a sense that's blocked with a stuffy nose.

TOUCH

How do we feel how soft cotton is or how cold ice cubes are? The **skin** is the **organ** responsible for the sense of **touch**.

The skin is divided into three layers:

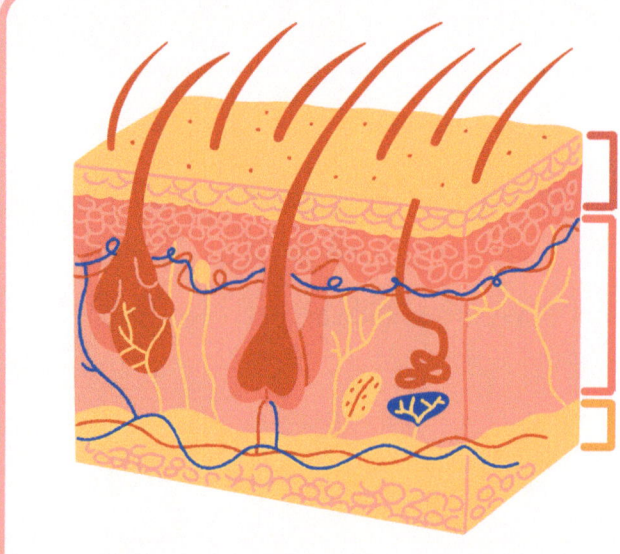

EPIDERMIS
Protects against fungi, bacteria, and viruses

DERMIS
Contains capillaries, sweat glands, hair follicles, and receptors

HYPODERMIS
Helps regulate body temperature, connects the skin to muscles

Receptors in the skin include **mechanoreceptors** (detect physical change), **thermoreceptors** (detect temperature), and **nociceptors** (detect pain). All of the receptors are connected to nerve endings, which send messages to the brain.

DID YOU KNOW?
Touch is the first sense to develop, starting around 7 weeks of pregnancy.

HOMEOSTASIS

Homeostasis refers to the body's ability to maintain a **stable internal environment** despite changes in external conditions. This is to ensure that all the bodily processes can occur properly.

For example, our body tries to maintain a **temperature** of around **37°C (98.6°F)**. This is the optimal temperature for a lot of chemical reactions that occur in our body.

To maintain this temperature, the body uses systems called **feedback loops**.

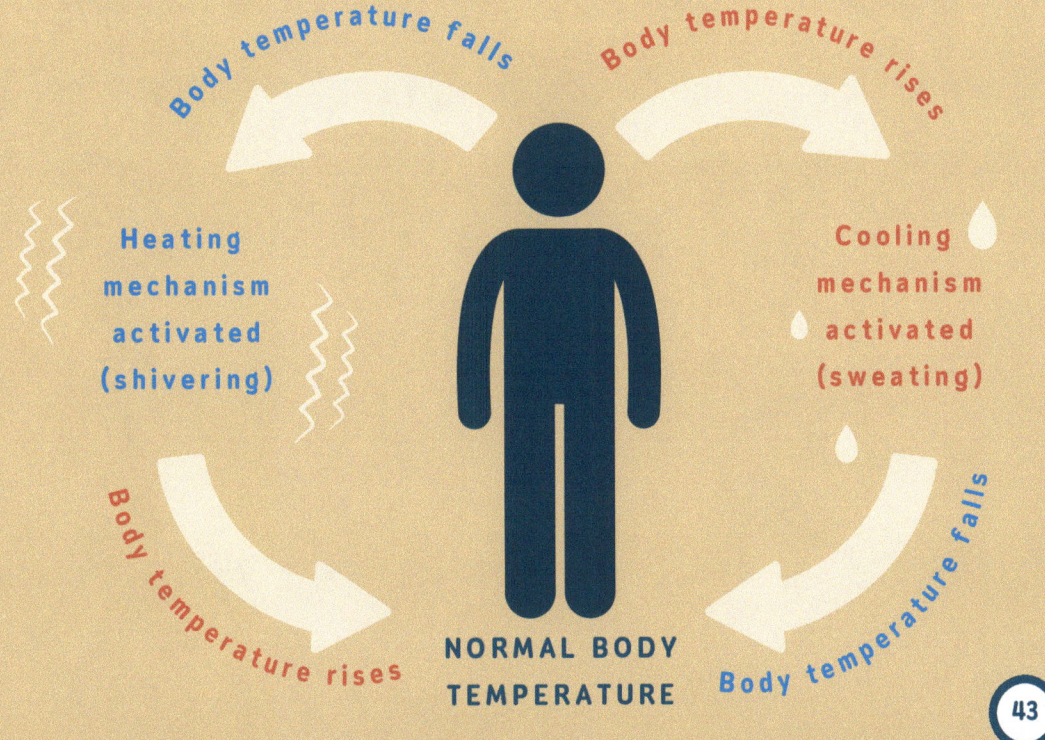

DISEASE

You can't catch a cold just by being out in the cold. Rather, **infectious diseases** are caused by **germs**, scientifically known as **pathogens**.

Pathogens are mostly microscopic organisms, and they include:

BACTERIA
E.g. Tuberculosis is a lung infection caused by the bacterium *Mycobacterium tuberculosis*.

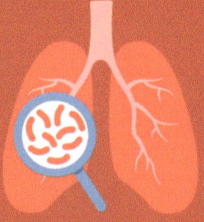

VIRUS
E.g. Measles is caused by the virus Measles morbillivirus.

FUNGI
E.g. Ringworm is skin infection caused by a group of fungi called dermatophytes.

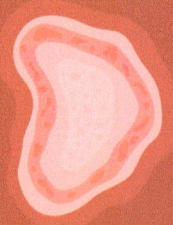

PARASITES
E.g. Malaria is a disease spread by parasite-infected mosquitoes.

Not all viruses, bacteria, parasites, and fungi are harmful to humans. For example, earlier, we saw how **bacteria** in our **digestive system** help us break down food.

 You can get infectious diseases by coming into contact with pathogens, such as by drinking contaminated water or shaking hands.

Some infectious diseases are not contagious from one human to another. Instead, they spread from animals or insects. For example, you can't catch **Lyme disease** from someone you pass on the street; it comes from the bite of an **infected tick**.

> People often say to stay warm to avoid catching a cold because some **viruses** multiply more effectively in cooler temperatures, and your **immune system** might not work as well when you're cold (see the next page for more info on your immune system).

Some diseases are not infectious. Examples of non-infectious diseases include **Alzheimer's disease**, which is associated with memory loss and cognitive decline; **Type 2 diabetes**, which can be triggered by lifestyle factors; **sickle-cell anaemia**, an inherited disease where blood cells are abnormally shaped; and **autoimmune diseases**, where the body's immune system mistakenly attacks its own tissues.

IMMUNE SYSTEM

You may not know it, but your body is engaged in a never-ending battle. The **immune system** is an army of **cells, tissues, and organs** that fight off harmful **pathogens** trying to infect you.

The immune system can be divided into **three lines of defence**.

FIRST LINE OF DEFENCE

These are **barriers** that are **non-specific**, meaning they do not target a specific pathogen.

Bacteria

Skin

Physical barriers provide a basic defence against invading pathogens. These include the skin, mucus, eyelashes, and eyelids.

The first line of defence also includes **chemical barriers**, such as the acidic fluids in urine, which can destroy pathogens. Compounds found in tears, sweat, and saliva are also antimicrobial, meaning they can destroy pathogens.

SECOND LINE OF DEFENCE

These defences are also **non-specific**, but they are **cellular** and **molecular** responses to pathogens.

In the second line of defence, **white blood cells** called **phagocytes** engulf infected cells and foreign bodies. The site becomes **inflamed** to increase **blood flow** to the area and bring more immune cells to fight the infection.

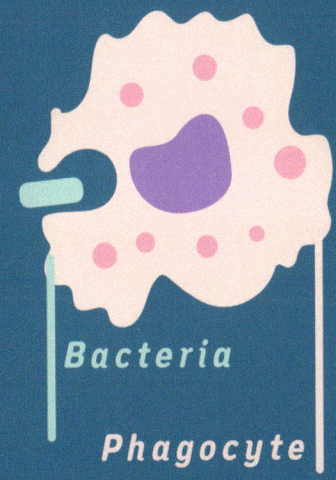
Bacteria
Phagocyte

THIRD LINE OF DEFENCE

When pathogens overcome the first two lines of defence, the **adaptive immune system** comes into play. These defences target **specific** pathogens and are **acquired** over time. It does so using white blood cells called **lymphocytes**. These can recognise **antigens** – molecules found on the surfaces of **pathogens**. They produce **antibodies** that bind to these antigens, tagging them for destruction by **phagocytes**.

Antibody
Antigens
Pathogen

Some lymphocytes become **memory cells**, which remain in the body after the infection to help your body respond quickly if the same pathogen returns. This is why we typically get certain illnesses, like chickenpox, just once in our lives.

VACCINES

Most people get many vaccines in their lifetime to protect against certain illnesses, but how do they work? Vaccines prepare our **adaptive immune system** to detect and attack **pathogens** by safely introducing antigens that it can "train" on.

Some vaccines, called **live attenuated vaccines**, introduce **weakened** versions of pathogens. These are generally not strong enough to cause disease. However, like the disease, they trigger an **immune response**, leaving behind **memory cells** that become useful if the same pathogen invades the body again.

Others are called **subunit vaccines**. These contain **parts** of a pathogen. For example, the vaccine for human papillomavirus (HPV) uses hollow, virus-like particles made from a protein found in HPV.

> Vaccines also protect people who cannot get vaccinated due to **herd immunity**. This is when a large percentage of the population is immune to a disease, reducing the likelihood of pathogens reaching vulnerable individuals.

ANTIBIOTICS

While vaccines are preventative, antibiotics can be used to treat **bacterial infections** after they occur by **killing** or **inhibiting the growth** of **bacteria**. For example, **penicillin** kills bacteria by destroying their **cell walls**.

> Sometimes people take antibiotics for illnesses caused by a **virus**, but antibiotics have no effect on viruses.

Sometimes, antibiotics may not be able to kill *all* bacteria. The toughest may have **resistance** against an antibiotic and **survive**.

These **multiply**, passing on and spreading the resistance, until the antibiotic is no longer effective.

This is why you should only take antibiotics when you absolutely need to.

WHAT ARE PLANTS?

We've explored a lot about the human body so far. Now, let's shift our focus to plants. Plants are living organisms that **photosynthesise**, meaning that they can make their own food using sunlight (see the page to your right to learn more).

TYPES OF PLANTS

Some plants are **vascular plants**. This means they have vessels called the **xylem**, which transports **water**, and the **phloem**, which transports **food** in the form of glucose (a type of sugar). Most organisms you recognise as plants - trees, shrubs, and flower - belong to this group.

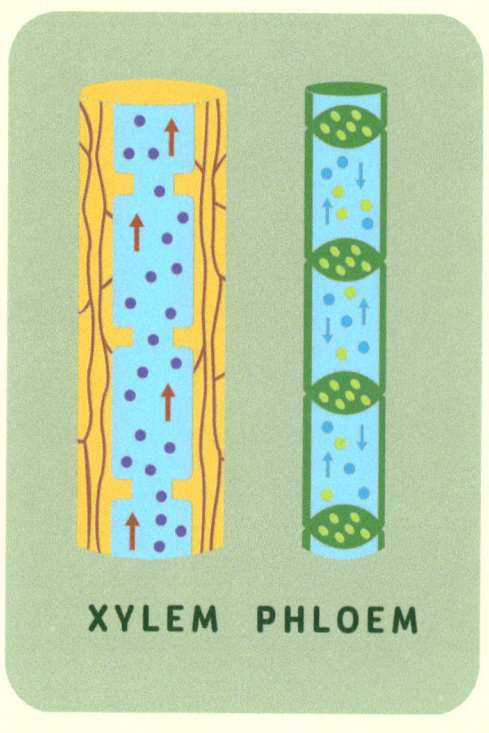

XYLEM   PHLOEM

If a plant is **non-vascular**, it does not have vessels. However, it still needs to transport water and food. Instead, they use processes called **diffusion** and **osmosis**, which work like a sponge. These plants are generally smaller and include mosses and liverworts.

PHOTOSYNTHESIS

Plants can't eat food the same way we do. Instead, they get energy through a process called **photosynthesis**.

Plants exchange gases like we do, except they take in **carbon dioxide** and release **oxygen**.

Carbon dioxide from the air and **water** absorbed from the roots are reacted, with the help of sunlight and **chlorophyll**. Chlorophyll is the **green pigment** that makes leaves green. This reaction produces **oxygen**, which is released into the air, and **glucose**, a form of sugar that is used as food.

$$\text{carbon dioxide} + \text{water} \xrightarrow[\text{chlorophyll}]{\text{sunlight}} \text{glucose} + \text{oxygen}$$

DID YOU KNOW?
Certain types of algae and bacteria also use **photosynthesis**. All organisms that use photosynthesis are called **photoautotrophs**.

PLANT STRUCTURE

Most **vascular** plants have three basic parts:

> **LEAF**
> Site of **photosynthesis**; usually flat and thin to maximise sunlight capture

> **STEM**
> Supports leaves and flowers; contains **xylem** and **phloem** for transport.

> **ROOTS**
> **Anchor** the plant and take in water and minerals from the soil.

Root cells often have **root hairs**. This increases the **surface area**, allowing the plant to absorb more **water and nutrients**.

LEAF STRUCTURE

A leaf's structure is optimised so that it can conduct photosynthesis efficiently and ensure the plant's survival.

WAXY CUTICLE reduces evaporation (water loss)

PALISADE MESOPHYLL contains **chloroplasts**, which contain **chlorophyll** for **photosynthesis**

XYLEM and PHLOEM

DID YOU KNOW?
Autumn leaves are brown because the **chlorophyll** breaks down in autumn.

STOMATA
Stomata are openings made by two **guard cells**. They **open** and **close** to allow **gas exchange** and regulate **water loss**.

Closed

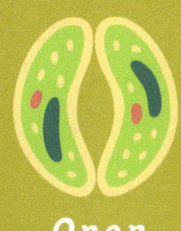
Open

Carbon dioxide enters through the stomata, and oxygen exits.

53

POLLINATION

Flowering plants reproduce through a process called **pollination**.

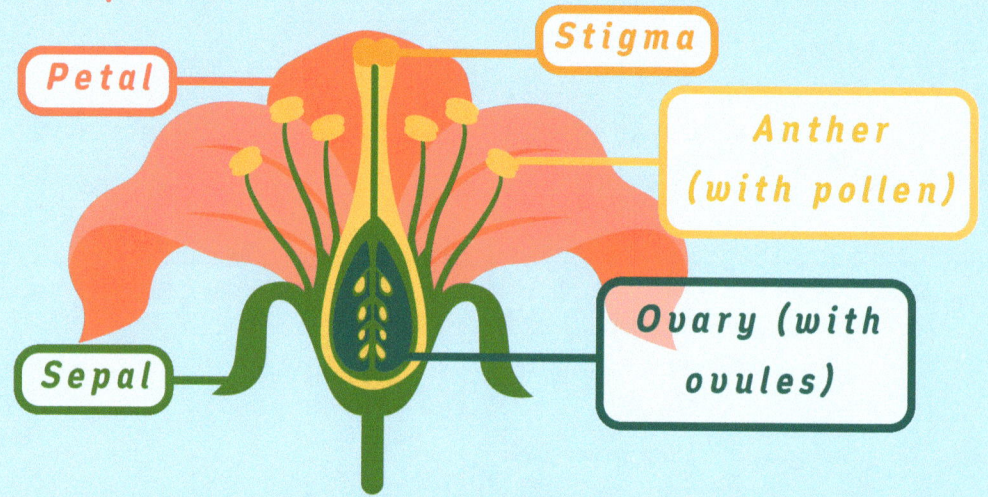

Pollination is when the **pollen** on the **anthers** is transferred to the **stigma**. This can happen directly through humans in a process called artificial pollination. It can also happen naturally by abiotic (non-living) vectors like the wind and water, or by biotic vectors like insects and birds.

Plants have brightly coloured **petals** and **nectar** to attract insects and birds. When the animals land, pollen sticks to their bodies and rubs off onto the next flower, helping to pollinate it.

DID YOU KNOW?
Declines in bees have been reported in all regions around the world due to habitat loss, pesticide use, and climate change. Fewer bees mean fewer plants get pollinated, which can impact ecosystems and agriculture.

After pollination, the pollen travels down from the **stigma** to the **ovary**, where it combines with the **ovules**, which are like eggs. This then forms a **seed**, and the ovary develops into a **fruit** to protect the seed.

Some flowers, like avocados, have only one ovule in their ovary, so their fruit has just one seed. Others, like apples, have many seeds inside the fruit.

These fruits are often eaten by **animals**, and the seeds are passed out in their droppings, usually some distance from the parent plant.

The seeds then go through **germination**, growing into seedlings that eventually develop into new plants.

INHERITANCE

Heredity is the passing of **genetic traits**, like eye colour and height, from parents to children. Traits such as dyed hair and piercing are not inherited because they do not involve changes to DNA.

People once believed children inherited a blend of their parents' traits. However, scientists later discovered that some traits can skip generations or appear unexpectedly. Today, thanks to a scientist named **Gregor Mendel**, we now understand some patterns in how traits are inherited.

You inherit half your DNA from your **mother** and half from your **father**, meaning you have **two copies of each gene** – one from each parent.

Each gene has two **versions** called **alleles**. For example, the gene for freckles has one allele for "freckles" and another for "no freckles." Some alleles are **dominant**, like the "freckles" allele, while others are **recessive**, like the "no freckles" allele.

If both your genes have the allele for "freckles", you will also have freckles.

If one of your genes has the allele for "freckles" (**dominant**) and the other has the allele for "no freckles" (**recessive**), you will have freckles, because the dominant trait is expressed.

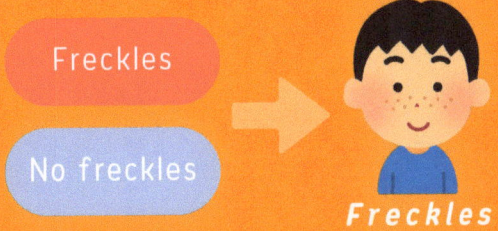

If both your genes have the allele for "no freckles" you will not have freckles.

BUT...

not all traits are passed on this way. Traits that follow this pattern are called "**Mendelian traits**". Freckles are a Mendelian trait. Having wet or dry earwax is also a Mendelian trait. Traits like height, hair colour, and skin colour are more **complex** and do not follow simple inheritance patterns. In fact, the way some genes work is still not fully understood.

ADAPTATIONS

Giraffes have long necks to reach leaves high in trees, and polar bears have thick fur to stay warm in the Arctic. These are called **adaptations** – traits that help species **survive** in their **environments**.

Adaptations can be divided into three types: **physical**, **physiological**, and **behavioural**.

PHYSICAL ADAPTATIONS

Physical adaptations are changes to an organism's body or **structural** features that help it survive in its environment.

- **Ducks** have **webbed feet** to make **swimming** easier.

- **Tigers** have **striped** coats for **camouflage**.

- **Eucalyptus trees** grow in Australia, where water is scarce. Their leaves **hang vertically** to minimise sun exposure to reduce **evaporation**.

PHYSIOLOGICAL ADAPTATIONS

Physiological adaptations are internal, biological changes that help organisms survive.

Snakes produce **venom** to paralyse **prey** and defend against **predators**.

Pigeons have **magnetite** in their bodies, which allows them to detect **magnetic fields** and orient themselves.

BEHAVIOURAL ADAPTATIONS

Behavioural adaptations are actions or responses that an organism exhibits to increase its chances of survival.

Whales migrate to warmer waters in winter, but return to nutrient-rich waters near the poles to feed.

Phototropism is when plants bend or grow towards the **sun**. A clear example is how young **sunflowers** bend to face the sun during the day.

Bears hibernate to slow down their body functions and save energy to survive the **cold winter**.

EVOLUTION

Evolution is the process by which all the kinds of living things that exist today developed from earlier forms. This idea contrasts with the old theory that species were created independently and remained unchanged.

The theory of evolution was first proposed by **Charles Darwin** in his 1859 publication, *On the Origin of Species*.

EVIDENCE FOR EVOLUTION

Darwin used **fossils** to show species that once existed but no longer do. Fossils also revealed gradual changes in organisms over time, indicating evolution had occurred.

He also discovered **homologous structures**, which are structures that share similar bone structures but different functions. For example, the human arm and the whale flipper are structurally similar. This suggests a common ancestry, indicating that species have evolved into their current forms. Darwin also highlighted the similarities between **embryos** of different animals, further indicating that species evolved from a shared origin.

Darwin also visited the **Galápagos Islands**, where he observed that the islands had many different species of finches. These species had different beak shapes, adapted to the type of food available on each island. For example, the ground finch has a large, strong beak to crack open hard seeds.

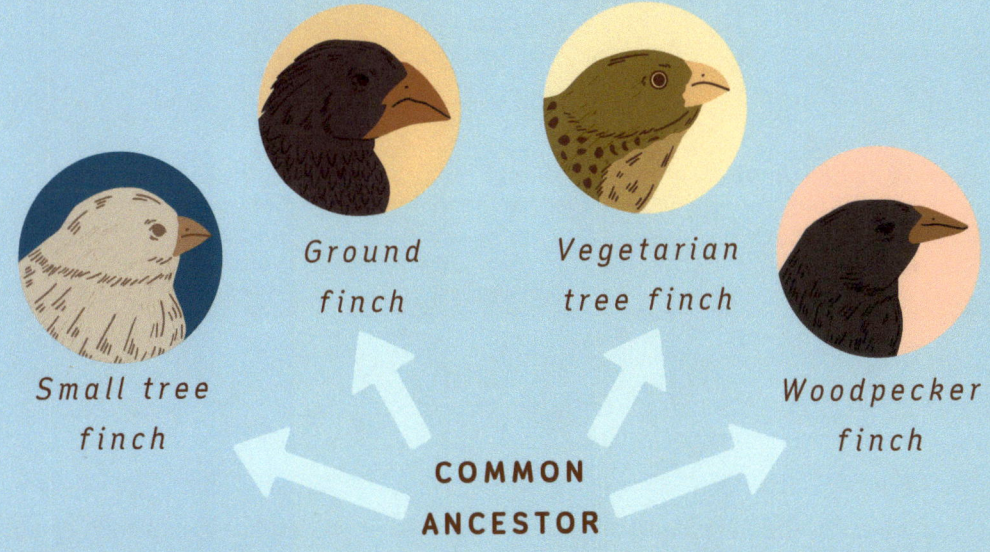

Small tree finch — *Ground finch* — *Vegetarian tree finch* — *Woodpecker finch*

COMMON ANCESTOR

NATURAL SELECTION

Based on his study of finches, Darwin proposed that **evolution** occurs through **natural selection**. This idea suggests that within a species, variations exist, and when there's **selection pressure**, individuals better **adapted** to the environment survive. For example, on an island with only seeds, finches with stronger beaks are more likely to survive and reproduce, making the strong beaks more common in the population. Over time, this can lead to the **evolution** of a new species.

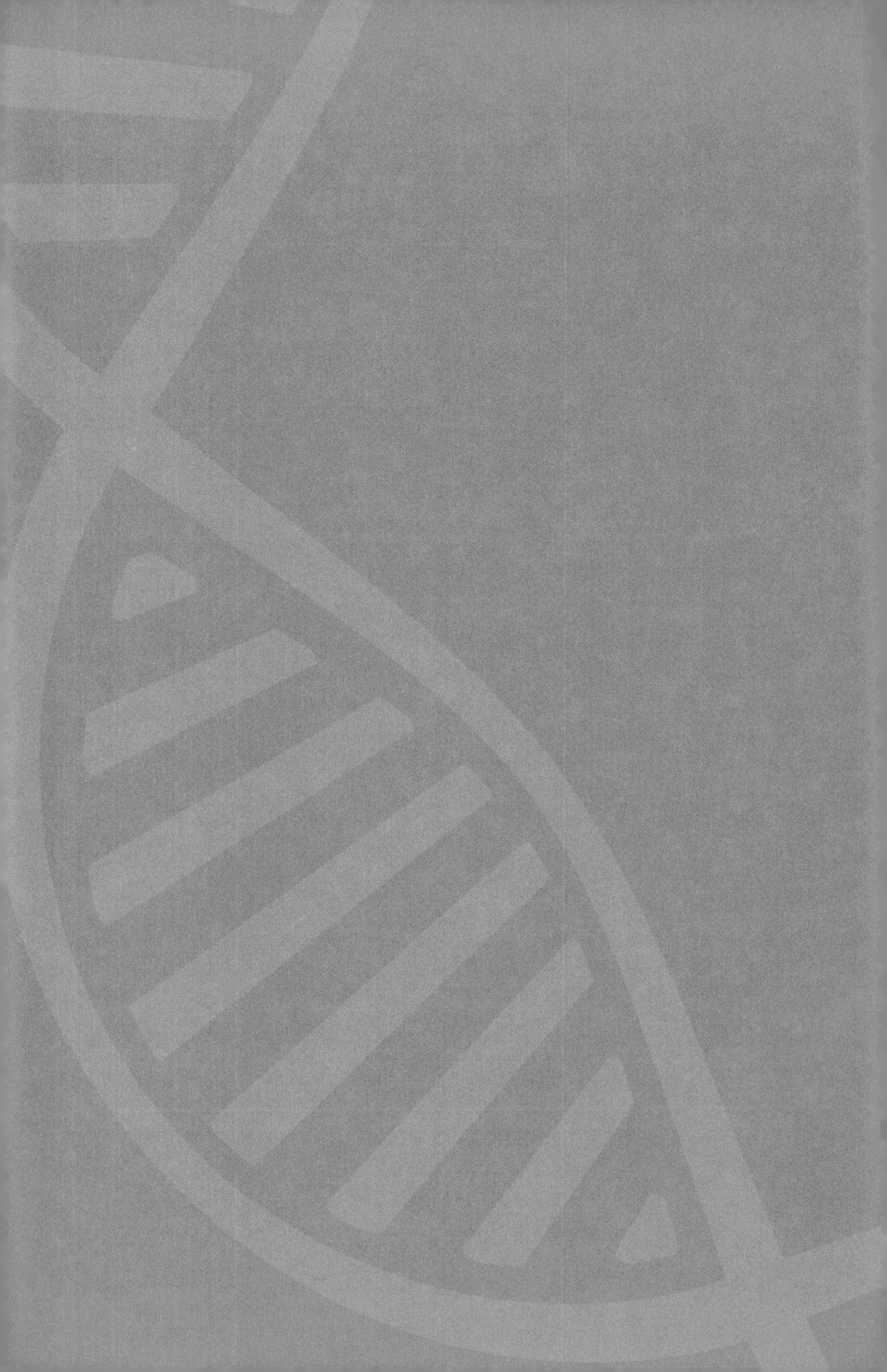

If you enjoyed reading this book, please leave a **review** on Amazon.

ACKNOWLEDGEMENTS

Thank you to everyone who provided invaluable feedback and advice throughout the development of this book. Your insights greatly enhanced its clarity and depth. A special thanks to expert consultant Jenny Zhang for your specialised knowledge and guidance in this field – your contributions were instrumental.

www.ingramcontent.com/pod-product-compliance
Lightning Source LLC
Chambersburg PA
CBHW041715160426
43209CB00018B/1837